Fairview Berries Presents

Berry Berry Sweet

Richard Rensberry

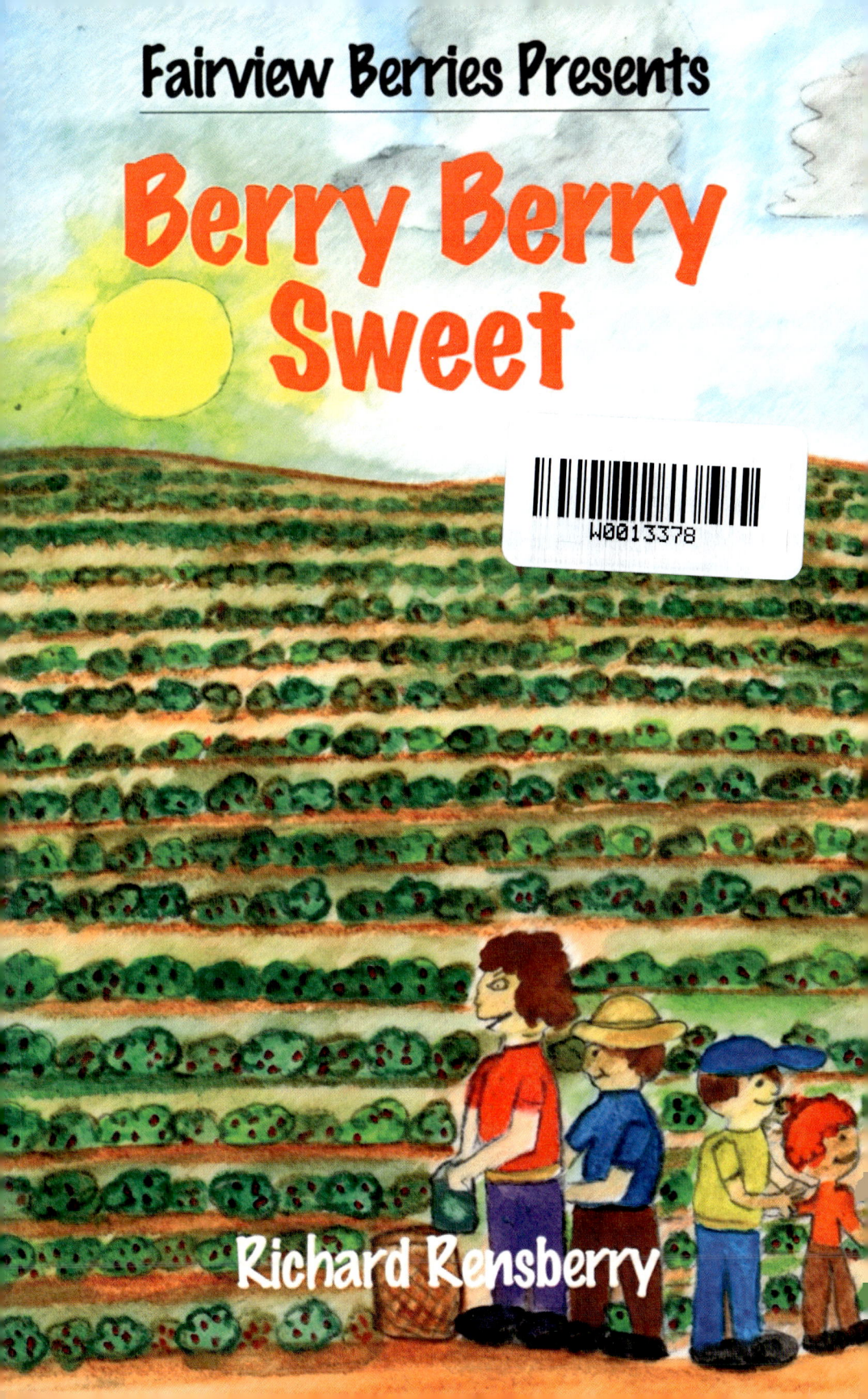

Published by: QuickTurtle Books LLC®

https://www.booksmakebooms.com

ISBN: 978-1-940736-66-2
Published in the United States of America

This book is dedicated to Fairview Berries of Fairview, Michigan. They grow strawberries, raspberries, grapes and other sweet things. They raise chickens, ducks and other animals. It is a magical farm full of kind and rambunctious good life.

THE HARDER
FARM

Berry Berry Sweet

The fields are groomed
for berry days
when pickers come
in long parades.

They kneel in rows
with hats pulled down
and sample taste
a berry found.

The roosters crow
to praise the sun.
The hens all cluck
and join the fun.

The dew is cool
with morning's blush.
The air is fresh
as bushes lush.

The pickers bend
with knees on mats,
fill cartons, bowls
and basket flats.

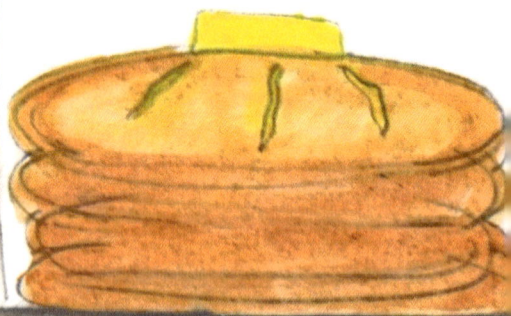

The berries ripe
are splendid red
for shortcake heaps
and jellies bled.

The fields are groomed
for berry days
when pickers come
in long parades.

They joke and laugh
at something said,
give hugs to friends
with lips stained red.

With baskets filled
for baking pies,
they pay their dues
and say good-byes.

The roosters crow
to praise the sun.
The hens all cluck
and join the fun.

Fairview Berries

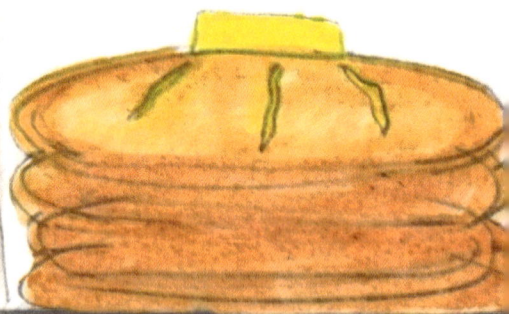

Strawberry Jelly

The End

Fairview Berries

Fairview Berries is a family owned producer of strawberries, grapes, chicken eggs and duck eggs for the local community of Fairview, Michigan. They have over 100 free range chickens and a flock of 17 ducks.

The very diverse variety of chickens on their farm lay multi-sized and multi-colored eggs. Each carton contains eggs that are green, blue, white and brown in hue. They are as beautiful as they are nutritious.

Emily Harder says they grew up in Detroit where the only chickens they had were fried. Not so in Fairview, here they have created a Fairyland where a walk through the yard is like walking into a book by Doctor Seuss. There are duck ponds, chicken runs, grapevines, strawberry fields, blackberry bushes, pumpkin patches, orchards and all the cacophony of farm life.

Fairview Berries is open for strawberry picking during the mid-summer months when they are also host to 4-H programs, Girl Scout activities and Exploration Club projects. They are located at 3788 Mast Road, Fairview, Michigan 48621.

Visit them at https://www.facebook.com/Fairviewberries/

Glossary

rambunctious- wild and noisy

groomed- taken care of

pickers- those gathering strawberries or other fruits

praise- express approval and admiration

blush- a rosy color

lush- thick and green

basket flats- wooden boxes with handles for carrying quart baskets of berries

splendid- very good, excellent

bled- taken from

dues- what it cost in money for the fruits they pick

Richard and Mary Rensberry are entrepreneurial authors who partner with and create children's and educational books for small businesses and worthy causes. Richard and Mary have written books for *The Wild Frontier Fun Park* of Comins, Michigan, *Michigan Lighthouses*, *Kenaf Partners USA,* of Onaway, Michigan, and *Artists For A Better World,* of Los Angeles, California. They have several other projects in the making and are striving continuously to help small businesses grow by creating books that will help personalize and generate far-reaching interest in each businesses' brand for generations to come. They can be reached at **maryandrichard@quickturtlebooks.com** or **https://www.booksmakebooms.com** for more information about their books and services.

More QuickTurtle Books®
by Richard and Mary Rensberry

Grandma's Quilt
If I Were A Lighthouse
Keepers of the Light
Big Ships
Goblin's Goop
The Best of Me from a-z
If I Were A Garden
Sasquatch
I Wish I Could
Twelve Months Make A Year
Kenaf, Seeds for Life
Kenaf, Seeding the World
Butterfly Stomach
Bzzzz
If I Were A Caterpillar
Maple Tree Elves
If I were A Blossom

QuickTurtle Books®
330 Schmid Road
Fairview, Michigan 48621

Made in the USA
Middletown, DE
18 August 2020